Bunny Maths 1 Workbook

This workbook is functional and designed to encourage pupils in area of number work, time, transformation, basic algebra, general numeracy skills and basic data handling techniques using mathematical concepts.

It is suitable for children between the ages of 5 and 7 and is the second edition of Preparation Workbook 1 Mathematics.

This is an excellent revision package adapted for every child. It is suitable for USA, UK, Canadian, Asian, the rest of Europe and African educational system.

Bunny Maths is fully loaded with answers, test paper and lots of practice.

I0489758

5 – 7 Years

To order this book, visit www.amazon.com or www.iconicconcepts.co.uk. For bulk order, contact Iconic Concepts Limited by email: edward@iconicconcepts.co.uk or edwardoranye@hotmail.com.

Contents

1.1   Addition up to 10

1   2 + 3 = ⬡

2   5 + 4 = ⬡

3   6 + 0 = ⬡

> Fill in the missing numbers from questions 1 to 8.

4   2 + 8 = ⬡

5   7 + 3 = ⬡

6   4 + ⬡ = 7

7   ⬡ + 3 = 6

8   2 + 5 = ⬡

1.2 Subtraction up to 10

Fill in the missing numbers from questions 1 to 14

1)　4 − 3 = ☐

2)　6 − 4 = ☐

3)　3 − 1 = ☐

4)　9 − 5 = ☐

5)　5 − 2 = ☐

6)　4 − ☐ = 1

7)　9 − ☐ = 5

8)　☐ − 4 = 3

9)　☐ − 6 = 2

10)　10 − 3 = ☐

11)　10 - 10 = ☐

12)　9 - 3 = ☐

13)　8 - 0 = ☐

14)　7 - 1 = ☐

1.3 Colouring activity 1

Work out the answers to each question and by using the key, colour each diagram on the right.

Key: **10 = Red**   **7 = Blue**   **4 = Green**
5 = Brown   **20 = purple**

1) $3 + 7 =$ 10

2) $7 - 3 =$

3) $13 + 7 =$

4) $12 + 8 =$

5) $10 - 6 =$

6) $5 + 0 =$

7) $2 + 5 =$

8) $9 - 4 =$

9) $13 - 8 =$

10) $0 + 7 =$

11) $4 + 6 =$

12) $17 - 13 =$

1.4 One and two times tables

Fill in the missing numbers

1)	1 x 1	=
2)	1 x 2	=
3)	1 x 3	=
4)	1 x 4	=
5)	1 x 5	=
6)	1 x 6	=
7)	1 x 7	=
8)	1 x 8	=
9)	1 x 9	=
10)	1 x 10	=

11)	2 x 1	=
12)	2 x 2	=
13)	2 x 3	=
14)	2 x 4	=
15)	2 x 5	=
16)	2 x 6	=
17)	2 x 7	=
18)	2 x 8	=
19)	2 x 9	=
20)	2 x 10	=

Work out the value of the numbers under the smiley faces.

21) 😊 x 3 = 6

22) 1 x 😊 = 2

23) 😊 x 7 = 14

24) 2 x 😊 = 10

25) 2 x 😊 = 8

26) 9 x 😊 = 18

27) 😊 x 7 = 7

28) 😊 x 1 = 20

1.5 Words and Numbers

Write these numbers in words.

1) 1 ☐
2) 5 ☐
3) 10 ☐
4) 17 ☐
5) 20 ☐

6) Using a ruler, join each number to its word.

5　　**7**　　**10**　　**15**　　**20**

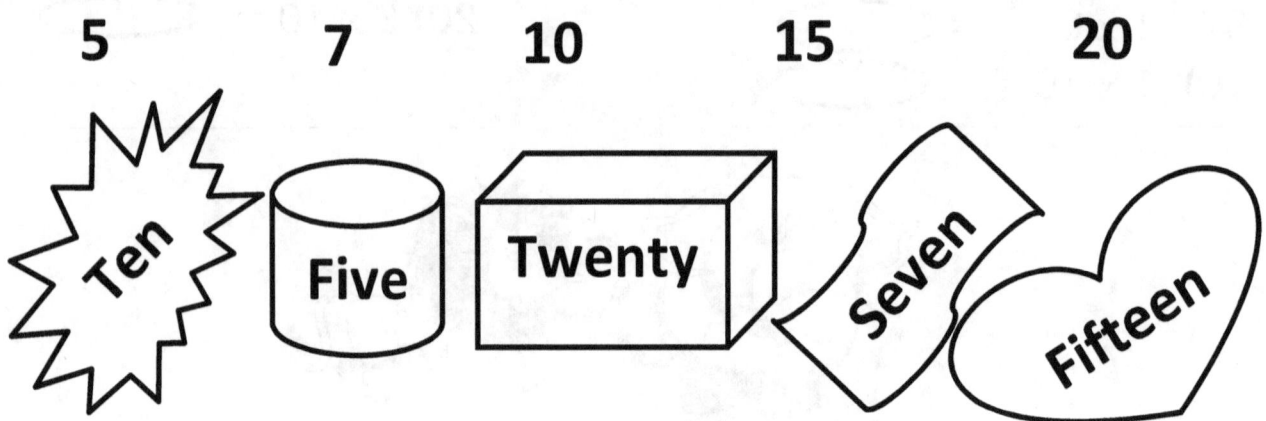

Ten　Five　Twenty　Seven　Fifteen

Write these words in numbers.

7) Eight ☐

8) Twelve ☐

9) Twenty ☐

10) Thirty ☐

11) Thirty – five ☐

12) Fifty – three ☐

13) Sixty ☐

14) sixty – six ☐

15) Seventy ☐

16) Seventy – nine ☐

17) Eighty ☐

18) Eighty – seven ☐

19) Ninety ☐

20) One hundred ☐

1.6 Small and big numbers

From the pair of numbers below, circle the **smaller** number.

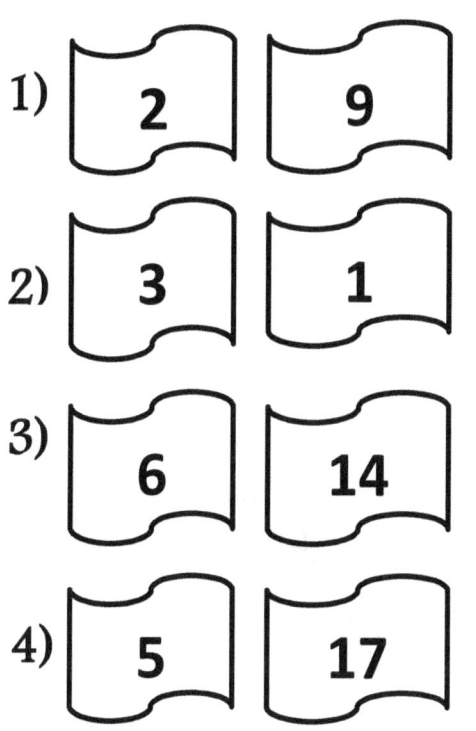

1) 2 9

2) 3 1

3) 6 14

4) 5 17

From the pair of numbers below, circle the **bigger** number.

5) 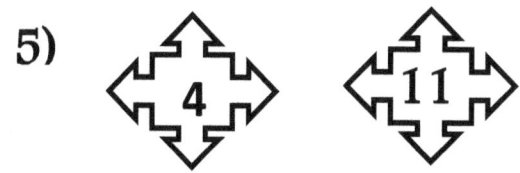 4 11

6) 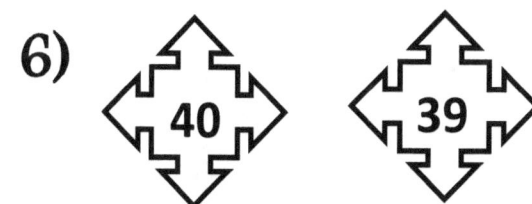 40 39

7) 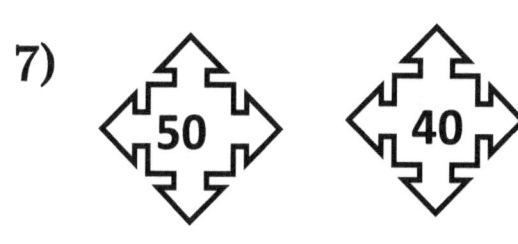 50 40

Put a tick on the **smaller** number from each pair.

8)  3.4 5.4

9) 8.1 8.0

10) 1.1 1.7

11) 12.5 21.5

12) 5.5 4.4

13) 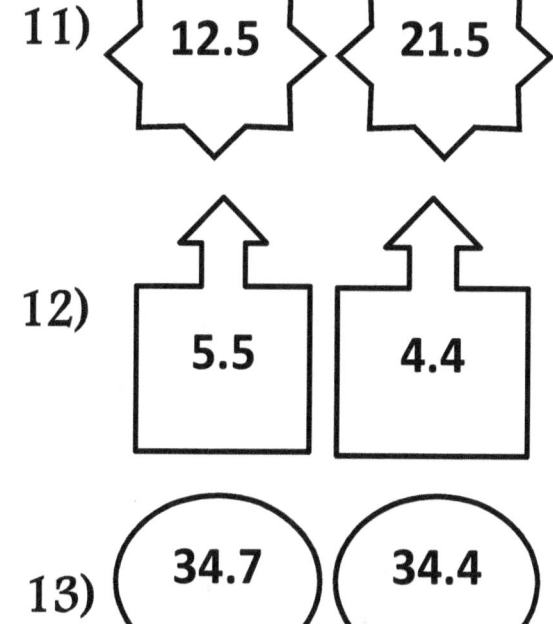 34.7 34.4

2.1 Adding smaller numbers

1 Add the following:

a) 12 + 8 = ☐

b) 11 + 9 = ☐

c) 7 + 18 = ☐

d) 12 + 0 = ☐

2 Add the following:.

a) 11 + 12 = ☐

b) 13 + 15 = ☐

c) 6 + 17 = ☐

d) 21 + 6 = ☐

3 Fill in the boxes below.

a) 2 + ☐ = 10

b) 4 + ☐ = 9

c) ☐ + 7 = 9

d) ☐ + 9 = 15

e) 0 + 11 = ☐

f) 9 + ☐ = 20

g) 20 + ☐ = 50

h) 40 + ☐ = 50

i) ☐ + 30 = 40

j) ☐ + 10 = 20

k) 0 + 50 = ☐

l) 5 + ☐ = 45

4 8 Penguins were found in a river. **12** more were added to the river. How many Penguins altogether are there in the river? ☐

5 Isabel wants to buy a ruler from a shop. The cost of one ruler is **20 pence**. She bought **3** rulers.
a) How much did Isabel pay for the three rulers? ☐
b) Isabel paid for the rulers with a **£1** coin.
How much change would she get? ☐

6 Sarah has **£5** for shopping. She wants to buy **2** calculators.
Does she have enough money? ☐

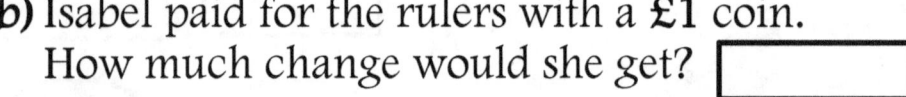
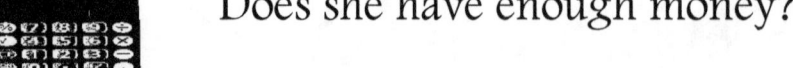

Price: £3

2.2 Even and Odd Numbers

Even numbers end in 0, 2, 4, 6, or 8. Examples: 2, 4, 6, 8, 10, 12, 14, 16...

1 From the numbers in the cloud, tick the even numbers.

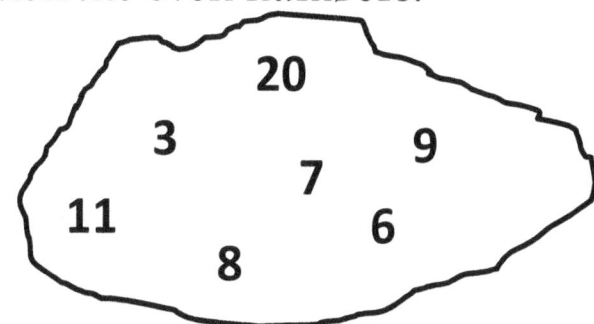

40 years Dad   Mom 35 years

**Billy
9 years**

2 From the list of numbers above, list all the odd numbers.

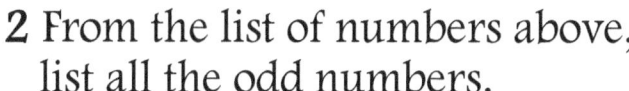

3 Write down all the even numbers between 19 to 30.

4 Look at the numbers below:

3, 5, 6, 10, 13, 20

a) Add all the even numbers.

b) Add all the odd numbers.

c) Write down the **difference** between the total of all the even numbers and the odd numbers.

5 Whose age is even?

6 Altogether Billy, Mom and Daddy's age is

7 The answer to question 6 above is even. True or false?

8 Dad is older than Billy by   years.

9 The answer to question 8 above is even. True or false?

10 Colour all even numbers **red** and all odd numbers **blue.**

1	2	3	4	5	6	7	8
9	10	11	12	13	14	15	16
17	18	19	20	21	22	23	24
25	26	27	28	29	30	31	32
33	34	35	36	37	38	39	40
41	42	43	44	45	46	47	48
49	50	51	52	53	54	55	56

2.3 Number Patterns (Sequences)

1) Fill in the missing numbers

   a) 1, 2, 3, 4, ☐ ☐

   b) 2, 4, 6, 8, ☐ ☐

   c) 5, 10, 15, 20, ☐ ☐

   d) 1, 3, 5, 7, 9, ☐ ☐

   e) 20, 18, 16, 14, ☐ ☐

2) Write the missing numbers for the sequences below.

   a) 4, 5, 6,_____,8,_____,10

   b) 10,_____, 30, 40,_____,60

   c) 7, 10,_____,16, 19,_____,25

   d) 1, 5, 9,_____,17, _____

   e) 22, 24, 26, 28,_____ , _____

☆ ☺ ⬭ ⬠ ✚ ⬛

| 6 | 10 | 14 | ☐ | ☐ | ☐ |

3) Fill in the missing boxes for the shapes numbers above.

4) Look at question 3 above. What word (Even or Odd) describes the numbers in the sequence? ☐

5) The rule for this sequence 3, 5, 7, 9, 11…….. is **ADD 2**
Write down the **rule** for each sequence below.

   a) 1, 4, 7, 10, 13,….. ☐

   b) 10, 12, 14, 16, …. ☐

   c) 20, 17, 14, 11, …. ☐

   d) 2, 7, 12, 17, ……. ☐

Number Patterns

6

O   OOO   OOOOO

Pattern Position: 1   2   3

    a) Draw the next pattern in the sequence. [   ]

    b) How many circles are added each time? [   ]

    c) Write down the **rule** for the pattern sequence. [   ]

7

Sequence	3	8	13		23	
Pattern position	1	2	3	4	5	6

    a) Complete the table above
    b) Write down the **rule** for the sequence [   ]

8 look at the number sequence below:
    40   35   30   25   20

    What is the rule for the sequence? [   ]

9 Write down the first **3** terms of the sequences below.

    a) First number is 1. Rule: add 2. [3,5,7]

    b) First number is 4. Rule: add 3. [   ]

    c) First number is 10. Rule: subtract 2. [   ]

    d) First number is 17. Rule: Subtract 3. [   ]

2.4 2-D Shapes

1) Colour the odd one out.

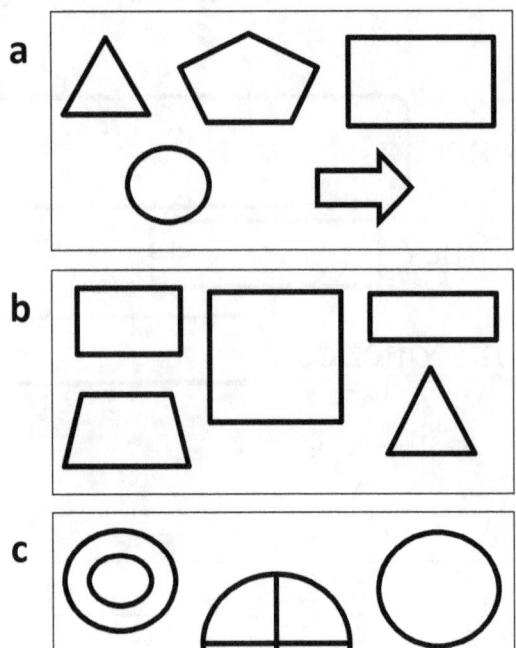

2) Write down the names of the shapes below.

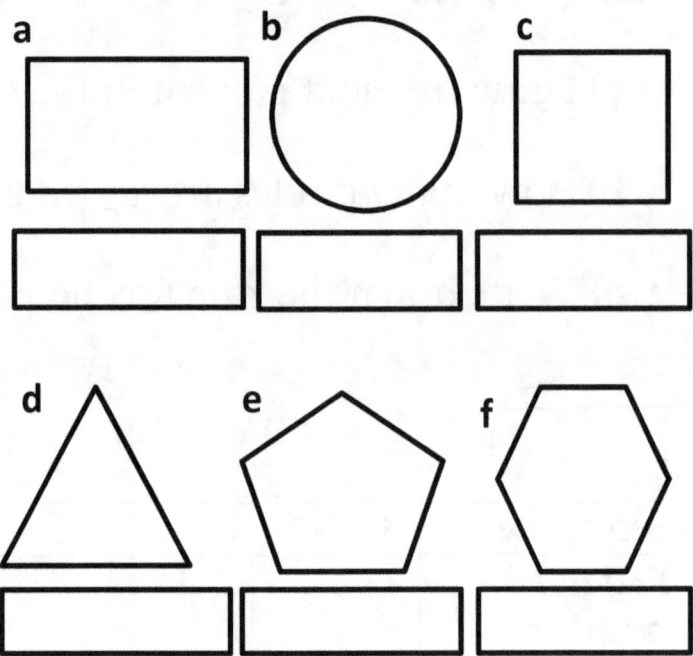

3) Names of 2D shapes are listed below.
   Join each of the shapes to their names.

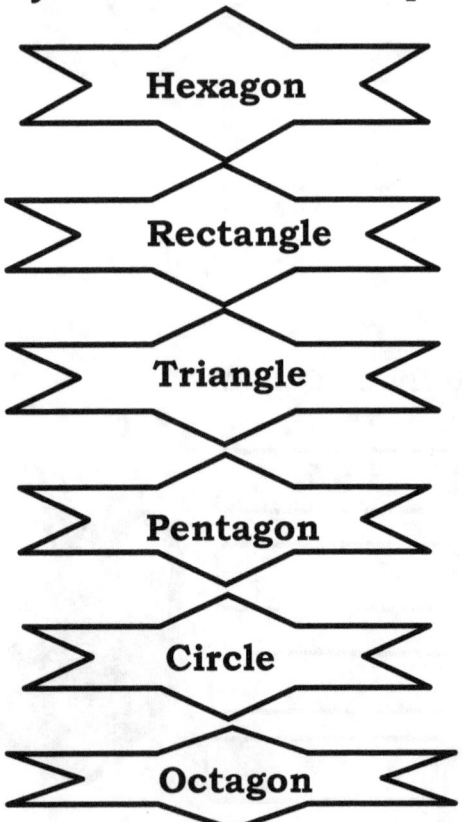

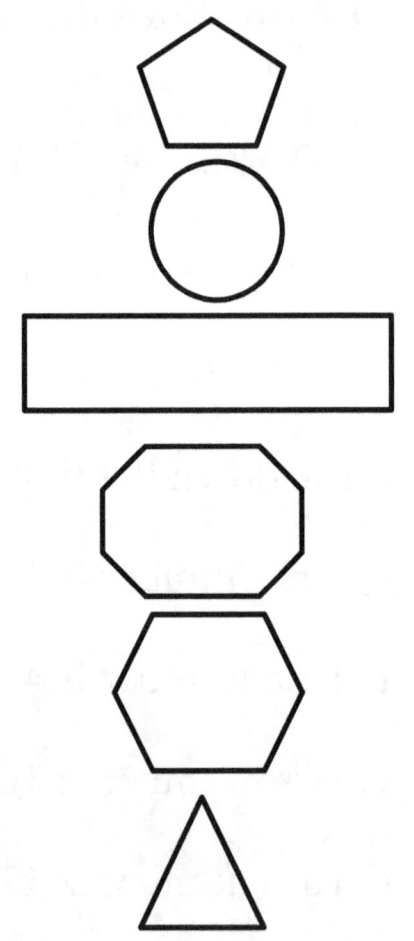

2.5 3-D shapes

1) Write the names of the 3-D shapes below.

a

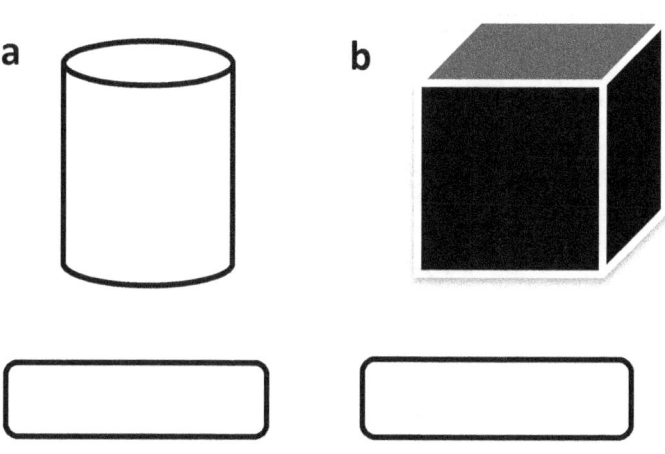

b

‎

a [_____] b [_____]

c

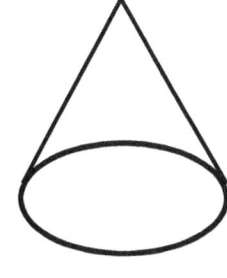

d

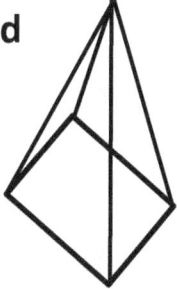

c [_____] d [_____]

2) Tick the odd one out.

a

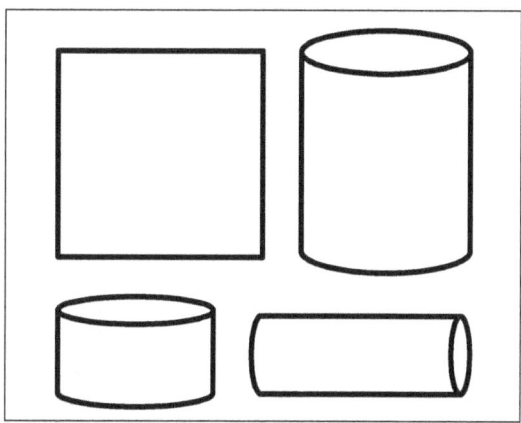

b

3) Ogugua says 'A sphere and a circle are the same'

Is Ogugua correct? [_____]

Explain your answer. [_____]

15

3.1 Symbols and basic equations

1) Work out the value of the **green** symbols in each equation.

a $3 + \boxed{} = 9$

$\boxed{} = \boxed{}$

b $10 + \boxed{} = 20$

$\boxed{} = \boxed{}$

c $1 + \boxed{} = 14$

$\boxed{} = \boxed{}$

d $\star + 7 = 20$

$\star = \boxed{}$

e $\star + 8 = 17$

$\star = \boxed{}$

f $\star + 25 = 50$

$\star = \boxed{}$

2 Write down the value of each symbol.

a $\boxed{} + 3 = 10$

$\boxed{} = \boxed{}$

b $\circledcirc + 6 = 6$

$\circledcirc = \boxed{}$

c $\boxed{+} + 5 = 20$

$\boxed{+} = \boxed{}$

d $\heartsuit - 5 = 15$

$\heartsuit = \boxed{}$

e $\square - 7 = 13$

$\square = \boxed{}$

f $\star + \star = 40$

$\star = \boxed{}$

g $\updownarrow + 100 = 150$

$\updownarrow = \boxed{}$

16

3.2 Thinking skills

1 Reece bought four cups for £2. Complete the sentences below.
a) One cup costs _____

b) Two cups costs _____

c) Eight cups costs _____

2 A ruler costs 70 pence. Isabel pays with a £1 coin.

a) How much change does she get? _____

b) James bought two rulers. How much did he spend?

c) Rosie had £3. She wants to buy 4 rulers. Does she have enough money? _____

3 Complete the missing numbers.
a) $2 + 7 = 1 +$ _____

b) $20 - 4 = 4 \times$ _____

PRICE: £25.56

4 Amber bought this picture and gave the shop assistant £25.

a) How much **more** will she pay? _____

b) Complete the boxes below for the amount of coins(pence) that will complete the payment

i) 20p + 20p + 10p + 5p + ☐

ii) 10p + 20p + ☐ + ☐ + 5p

iii) ☐ + 50p + ☐

c) Tom wants to buy the same picture. He gives £25.60. How much change will he get?_____

3.3 Addition Skills

Words used for addition:

Add

Plus

More than

Sum

Example   42 + 53

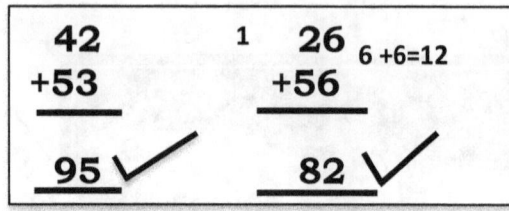

42	¹ 26 6 +6=12
+53	+56
95 ✓	82 ✓

1. Work out these additions.

a. 15 + 32   b. 24 + 31

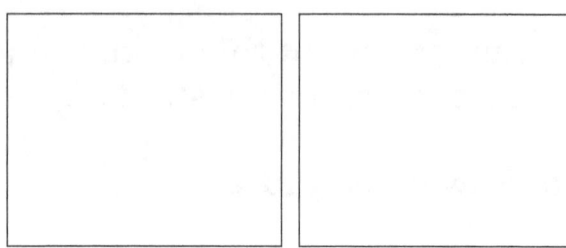

c. 23 + 53   d. 76 + 22

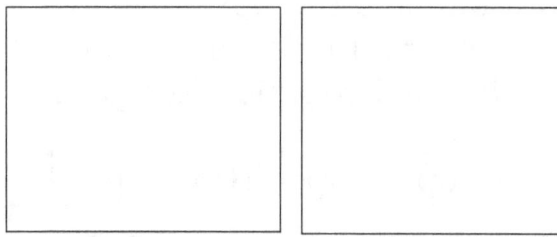

e. 55 + 42   f. 24 + 64

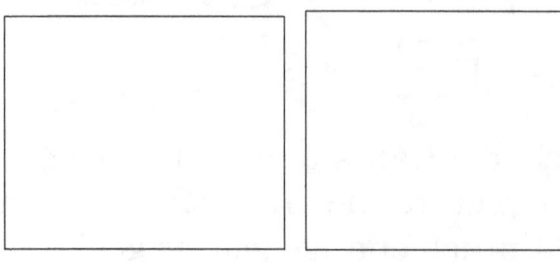

2. Work these out.

a. 13 +7 = …………

b. 30 + 50 = …………………..

c. 300 + 500 = …………………..

3. What is the sum of 33 and 15?

4. Add 12 and 83

5. 40 plus 50

6. What is 20 more than 200?

Add the numbers below:

7.   76 + 10 =

8.   48 + 30 =

9.   24 + 20 =

10. 71 + 23 =

11. 99 + 0 =

12. 36 +36 =

3.5 Subtraction Skills 1

35 – 13	94 - 43
$\begin{array}{r} 35 \\ -\,13 \\ \hline 22 \end{array}$ ✓	$\begin{array}{r} 94 \\ -\,43 \\ \hline 51 \end{array}$ ✓

Words used for subtraction.

Difference   Minu   Take away

Subtract these numbers.

1. 65 – 4

2. 29 - 7

3. 68 – 16

4. 44 - 10

5. 87 – 5

6. 71 - 30

7. Work out the difference between 16 and 6

8. 54 minus 44 =

9. 67 take away 13 =

Function Machines

Work out the missing numbers for 'IN' or 'OUT' in the function machines below.

10.   29 → -9 → OUT

11.   20 → -13 → OUT

12. IN → -3 → 5

4.1 Ordering whole numbers

Ordering simply means placing numbers in order of size from smallest to highest or highest to smallest.

4 is **bigger than** 2, 20 is **smaller than** 30 and 70 is **equal to** 70

< Represents smaller than
> Represents bigger than
= Represents equal to
Therefore, 2 is less than 3 and we may write as **2 < 3**
10 is greater than 7 and we may write as **10 > 7**

Write each number in order of size, smallest first.

1. 4, 3, 8, 6, 10, ..., ..., ..., ..., ...

2. 19, 10, 8, 3, 9, ..., ..., ..., ..., ...

3. 20, 10, 6, 4, 16, ..., ..., ..., ..., ...

4. 41, 14, 40, 56, 24,......., ..., ...,......., ...

5. 50, 56, 20, 76, 79, ..., ..., ..., ..., ...

Write the missing **signs** for each pair of numbers.

6. 5 $\boxed{<}$ 8   **7.** 30 $\boxed{\phantom{<}}$ 67

8. 10 $\boxed{\phantom{<}}$ 12   **9.** 80 $\boxed{\phantom{<}}$ 20

10. 20 $\boxed{\phantom{<}}$ 10   **11.** 21 $\boxed{\phantom{<}}$ 47

12. 7 $\boxed{\phantom{<}}$ 70   **13.** 200 $\boxed{\phantom{<}}$ 199

5.1 Time and Clock 1

This clock shows 2 o'clock

What time is shown on these clocks?

1 
2 
3 
4

5 
6 
7 
8

9 
10 
11 
12

13 
14 
15 
16

17 
18 
19 
20

21 
22 
23 
24

5.2 Time and Clock 2

1 Draw the hands of these clocks to show the times below each clock.

6 o'clock   **12 o'clock**   **9 o'clock**   **1 o'clock**

2 Write the missing clock numbers on the clock face below.

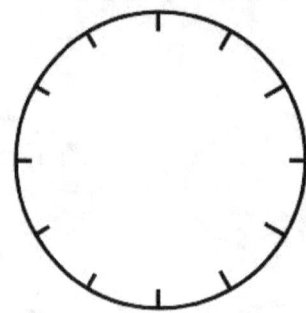

3 Draw the hands of these clocks to show the times below each clock.

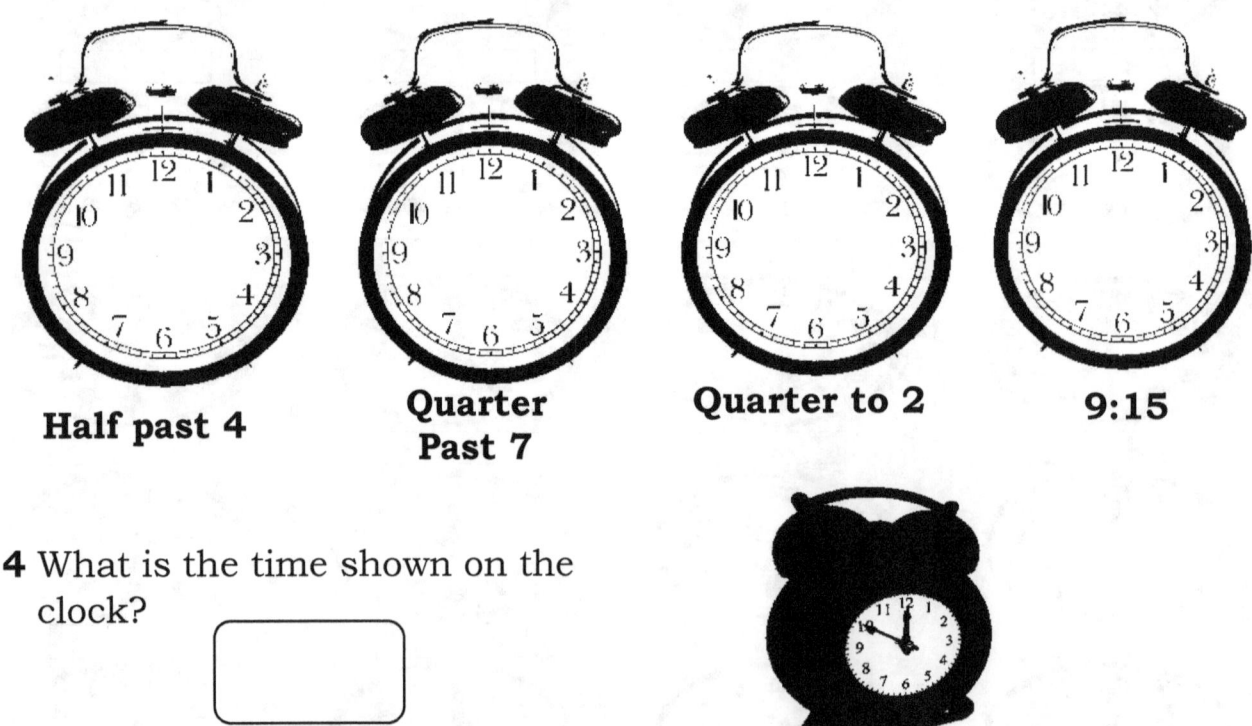

Half past 4   **Quarter Past 7**   **Quarter to 2**   **9:15**

4 What is the time shown on the clock?

6.1 Number pyramid

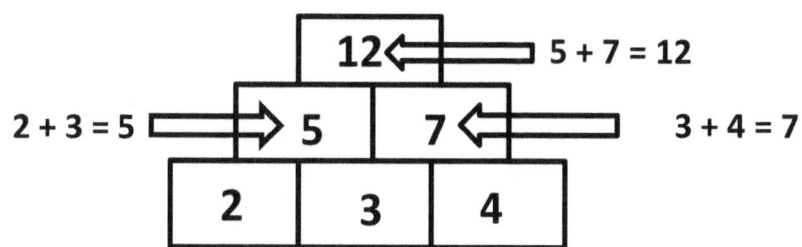

1) Look at the example above.
   Complete each pyramid below.

a
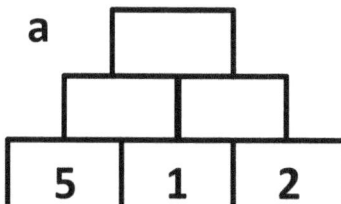

b
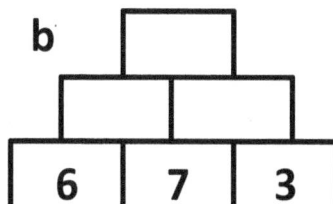

c
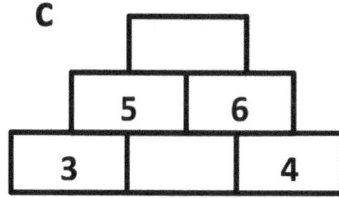

d
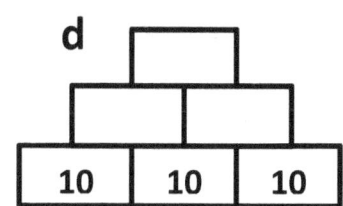

e
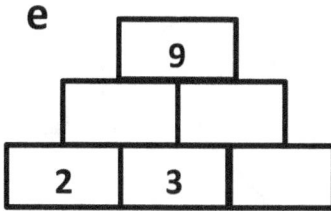

f
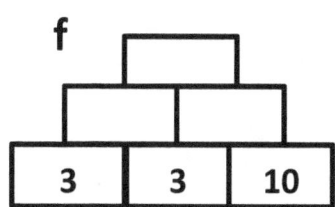

g
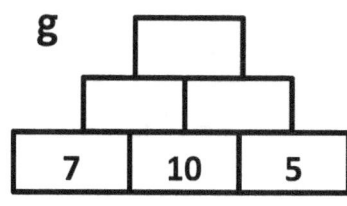

h
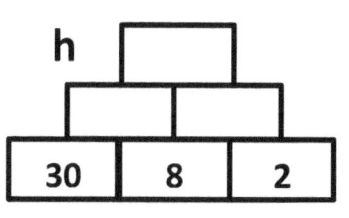

2) The diagram below is a **multiplication** pyramid. Complete the boxes.

a
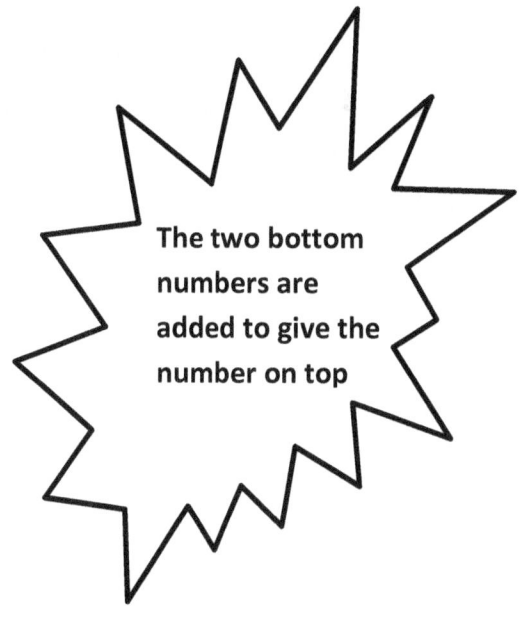

b
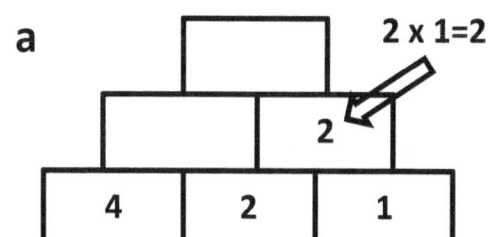

c
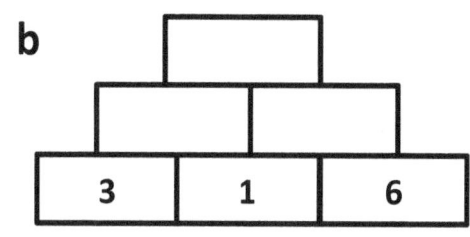
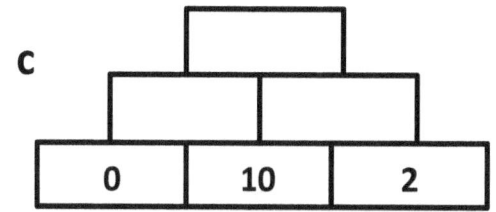

23

7.1 Symmetry

If we fold these shapes along the dotted line, one half of the shape will fit exactly on top of the other half.

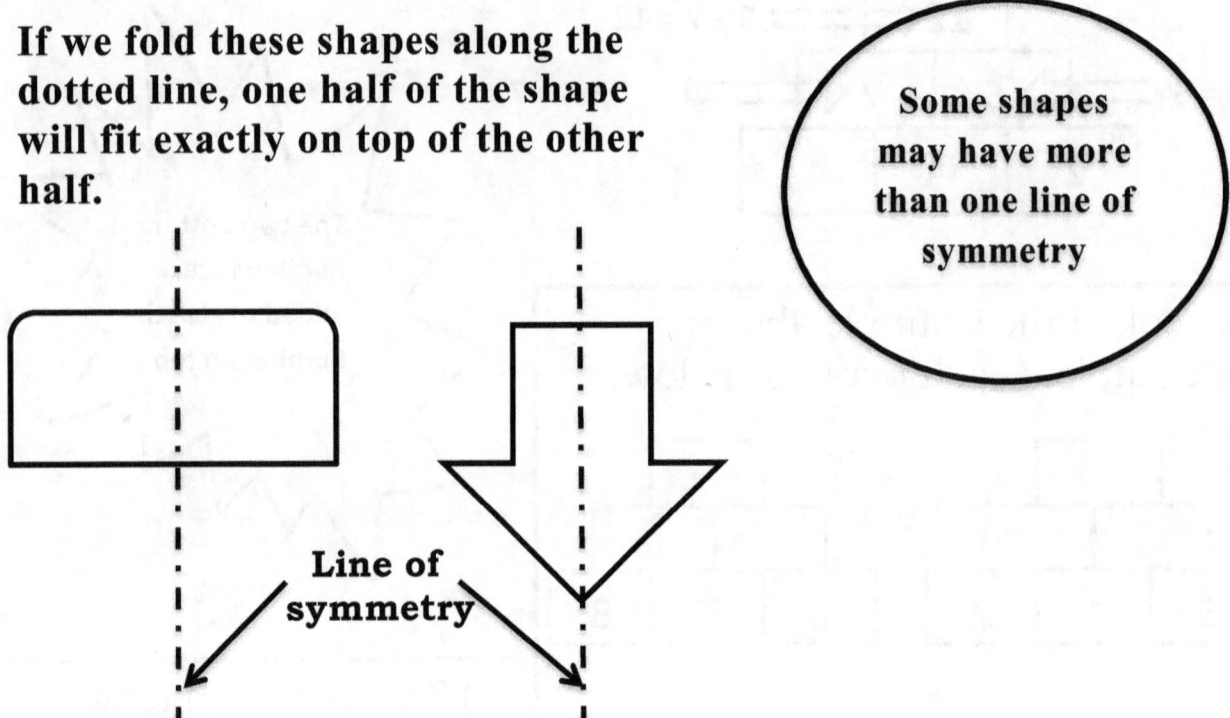

Some shapes may have more than one line of symmetry

Line of symmetry

The dotted line is called the **line of symmetry**. It is the line that divides the shape into **two** identical halves.

1 Complete the table and draw the line(s) of symmetry on each shape.

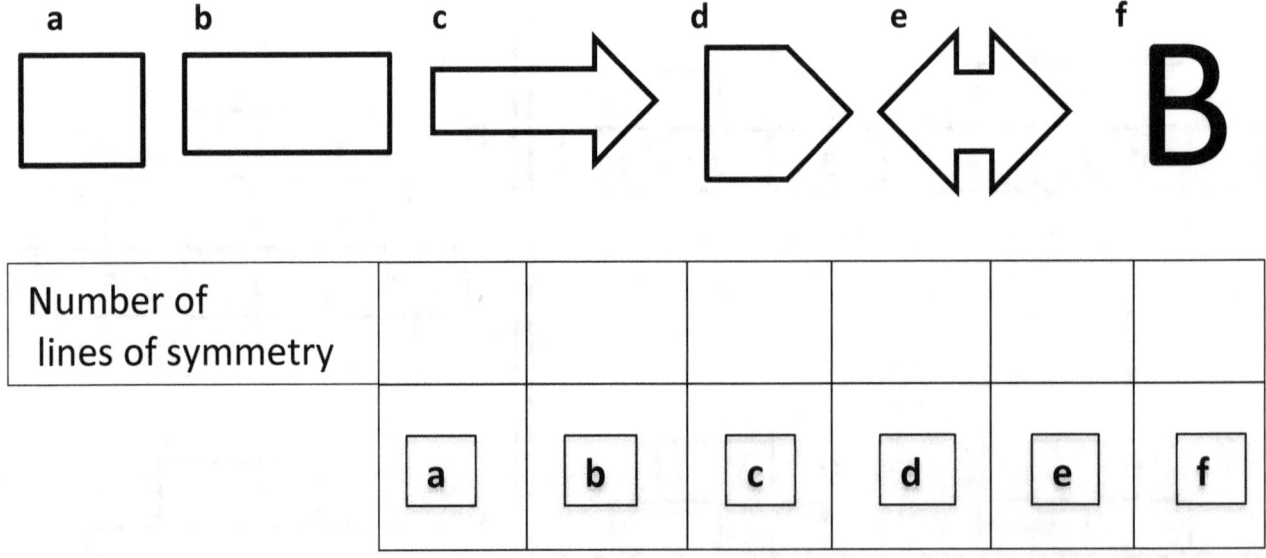

Number of lines of symmetry						
	a	b	c	d	e	f

7.2 Reflections

Can you draw a line of symmetry on the picture above?

1 Reflect (draw) each shape on the other side of the mirror line.

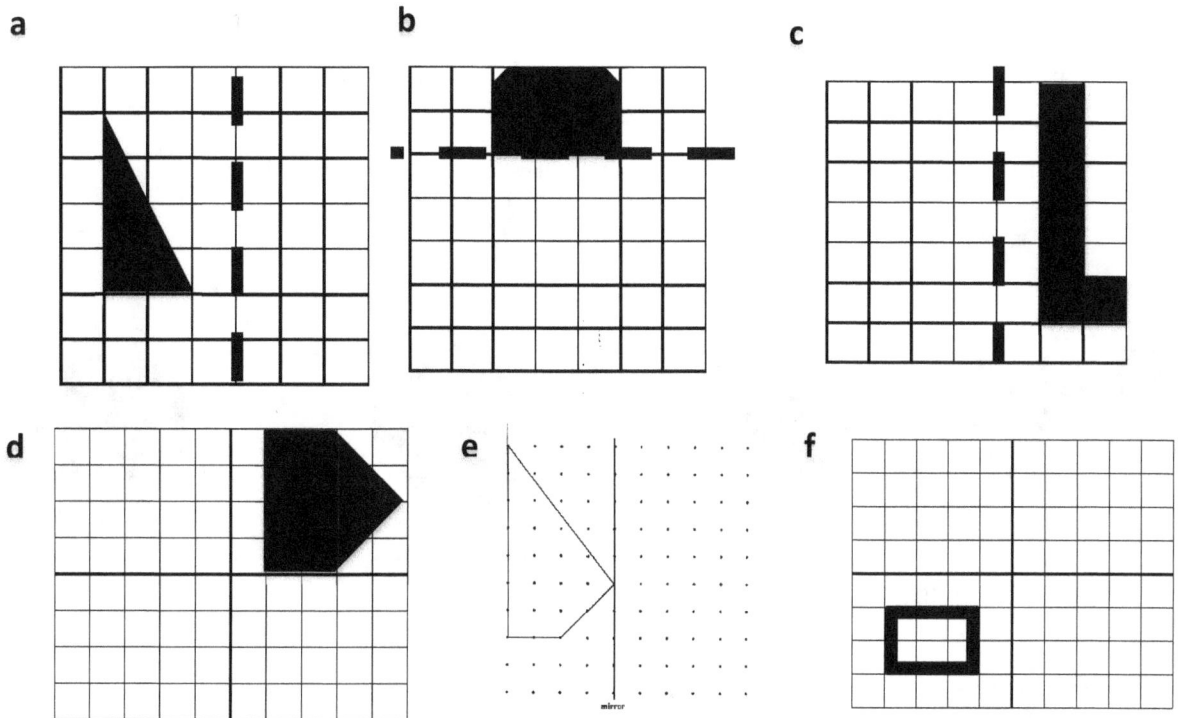

8.1 Dividing

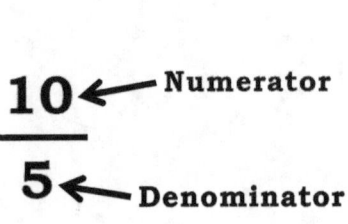

Numerator → 10
————
5 ← **Denominator**

3 x 5 = 15
15 ÷ 3 = 5
15 ÷ 5 = 3

Dividing simply means how many of the denominator can you get from the numerator.

$$\frac{8}{4} = 2 \checkmark$$ because 4 goes into 8 two times or 4 x 2 = 8

1 Write the missing numbers in the boxes provided.

a 8 ÷ 4 = ☐   **d** 27 ÷ 3 = ☐   **g** 9 ÷ 3 = ☐

b 10 ÷ 2 = ☐   **e** 18 ÷ 9 = ☐   **h** 25 ÷ 5 = ☐

c 18 ÷ 6 = ☐   **f** 4 ÷ 4 = ☐   **i** 20 ÷ 2 = ☐

14

18

24

20

2 a) Pineapple divided by **2** will give ☐

b) Strawberry divided by **9** will give ☐

c) Pumpkin divided by **8** will give ☐

d) Apple divided by **5** will give ☐

9.1 Colouring activity 2

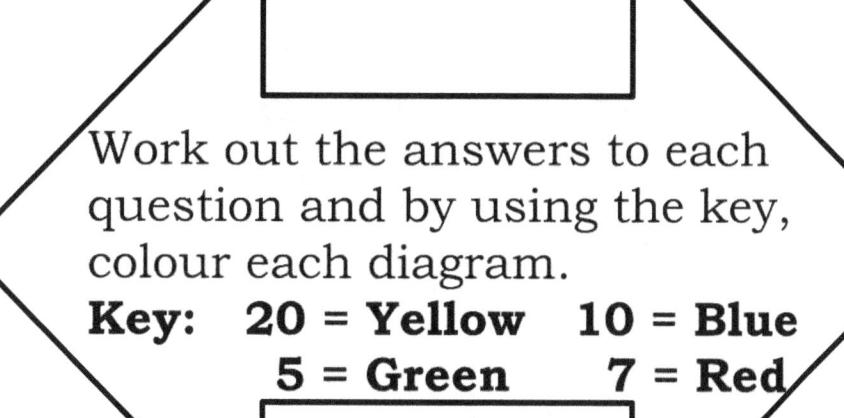

Work out the answers to each question and by using the key, colour each diagram.

Key: 20 = Yellow   10 = Blue
       5 = Green   7 = Red

1) $3 + 2 =$

2) $14 - 4 =$

3) $8 + 12 =$

4) $10 \div 2 =$

5) $4 \times 5 =$

6) $17 - 10 =$

7) $2 \times 5 =$

8) $25 - 5 =$

9) $10 \times 2 =$

10) $19 - 14 =$

10.1 Lengths and estimates

1) Estimate the length of each line and measure the exact length in cm.

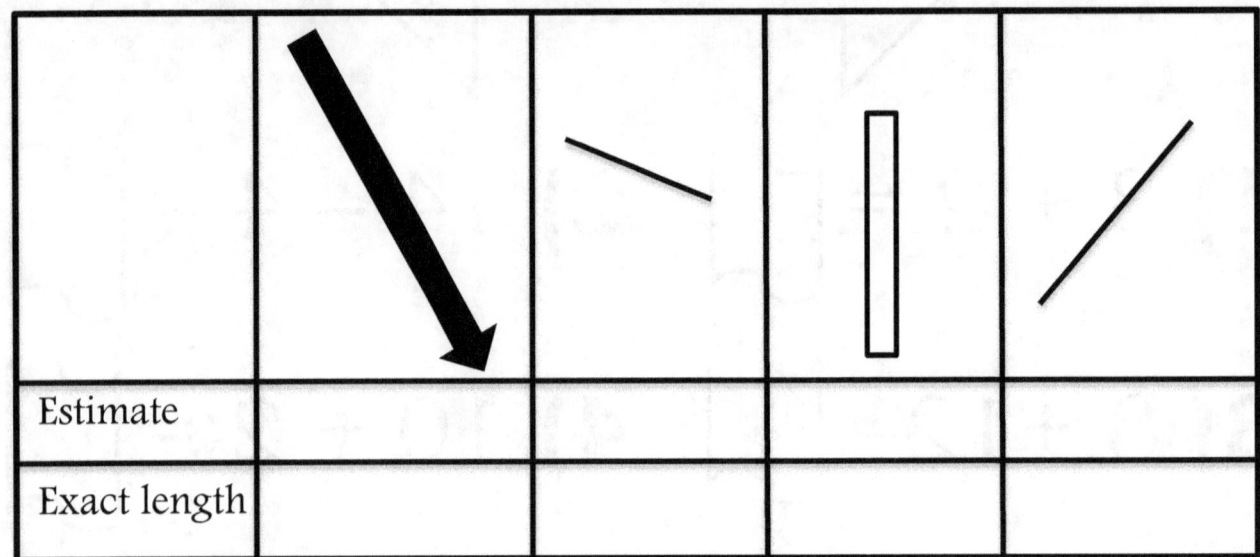

Estimate				
Exact length				

2) Measure and write down the length of each item in centimetre.

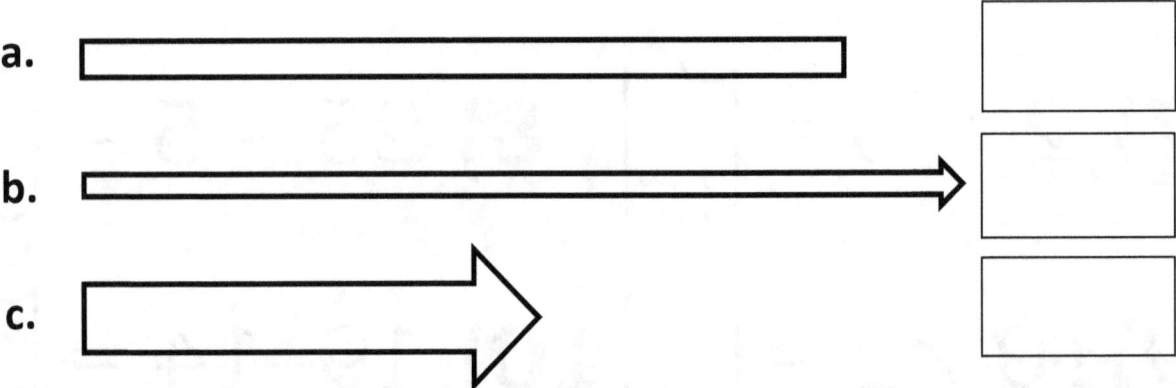

a.

b.

c.

3) Write down the measurements from question 2 above in millimetres.

10.2 Reading scale

1) What is the reading shown on the scale?

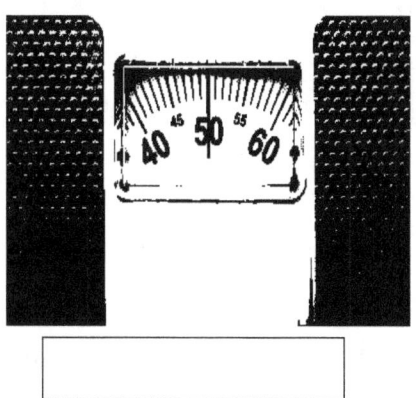

2) What is the reading on the scale?

3) Write down the readings on the scales below.

a

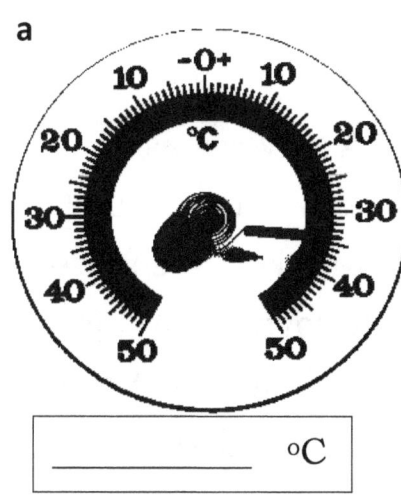

_____ °C

b

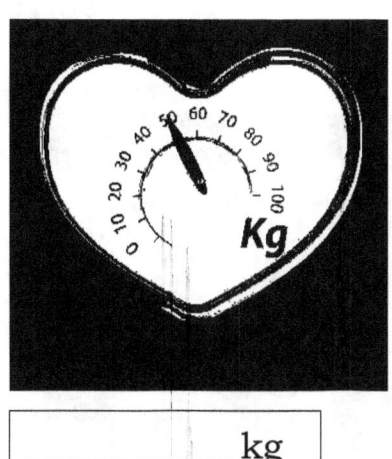

_____ kg

c

_____ kg

d

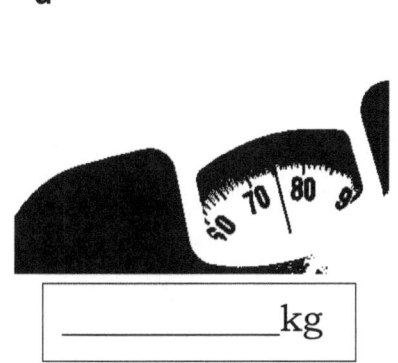

_____ kg

e

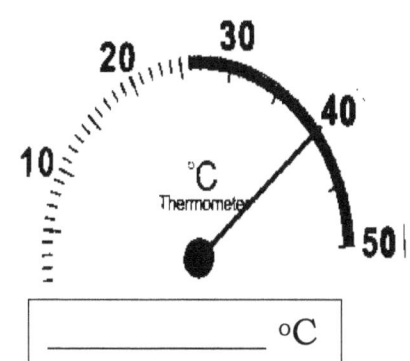

_____ °C

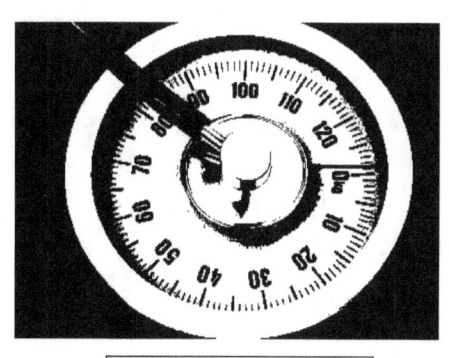

_____ kg

11.1 Frequency table and bar chart

Frequency table		
Colour of pencil	Tally	Frequency
Black		5
Grey	‖‖ ‖	
Blue	‖‖	

1) Some pupils were surveyed about the colour of pencils used in their classroom every morning. The result is shown on the frequency table.

a Complete the table.

b The favourite colour was ____

c How many pupils were surveyed? _____

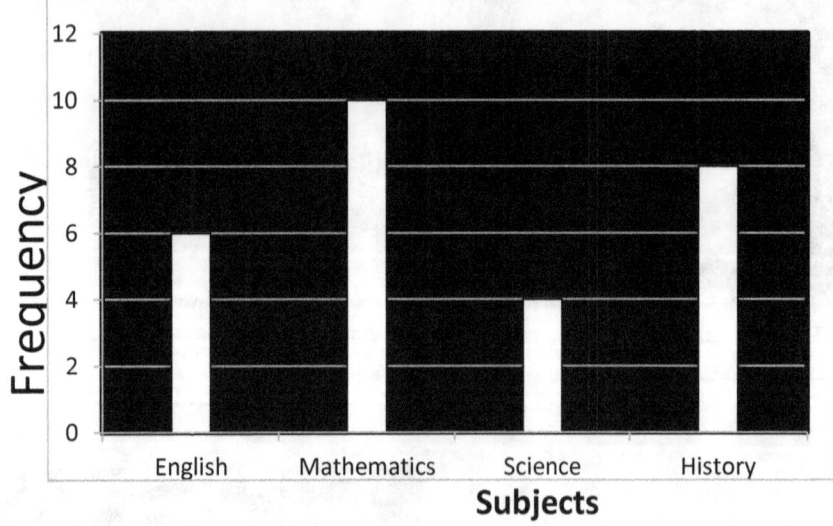

2) A survey was carried out and pupils were asked to choose their favourite subject. The results are shown in the bar chart.

a The most popular subject is ------------------------------------

b The least popular subject is ------------------------------------

c How many pupils chose English? ----------------------------

d How many pupils took part in the survey? ---------------

e Isabel says 'half of the pupils surveyed chose English and Mathematics' Is she correct? -----------------------------

TEST PAPER

* Calculator <u>not</u> allowed
* 50 minutes allowed
* 30 – 39 marks (Excellent)
* 20 – 29 marks (Good)
* 0 – 19 improvement needed
    GOOD LUCK!!!

5

Happie   Saddie   Bully

1 Put a tick on the smaller number from the pair of numbers below.

a)

| 32 | 23 |

b)

| 2.7 | 7.2 | 2 marks

2 Circle **all** the even numbers from the list below.

7   8   20   33   56

2 marks

3 Fill in the missing numbers.

a) $5 + \square = 9$

b) $1 \times 3 = 2 + \square$

c) $8 \div 2 = 2 \times \square$

d) $12 - \square = 4 + 6$

4 marks

4 Write the missing numbers for the patterns below.

a) 2, 4, ____, 8, 10, ____

b) 14, ____, 8, ____, 2

2 marks

a) Happie is **3kg** heavier than Saddie. How much does Happie weigh?

1 mark

b) Saddie, Happie and Bully the dog weighs 30kg altogether. What is the weight of Bully the dog?

2 marks

6 Some cylinders have **odd** and **even** answers. Colour the cylinders using the **key:**

Odd - Blue
Even - Green

3 +5

8 + 3

2 +12

7 + 0

2 x 7

3 x 3

5 + 5

3 + 12

12 - 6

9 marks

7 $14 + 19 = \square$

2 marks

8 From the number lines below, write the numbers marked with letters on the arrow.

a)

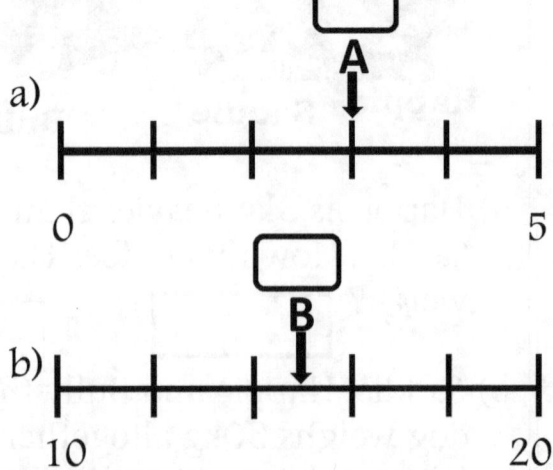

0   5

b)

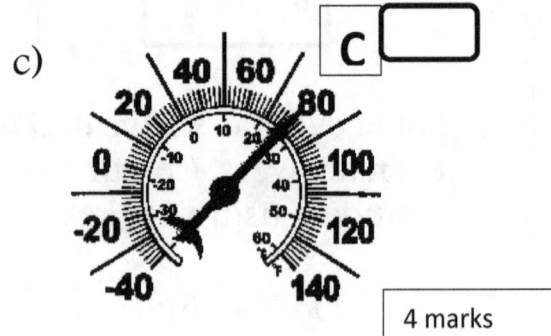

10   20

c)

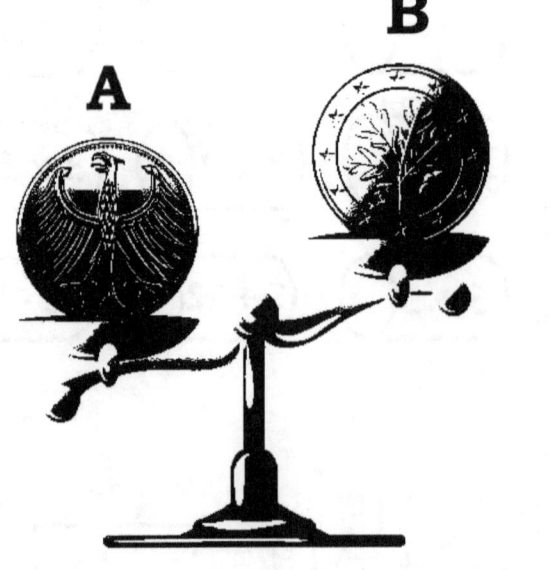

4 marks

9 Which coin is heavier, A or B? Put a tick.

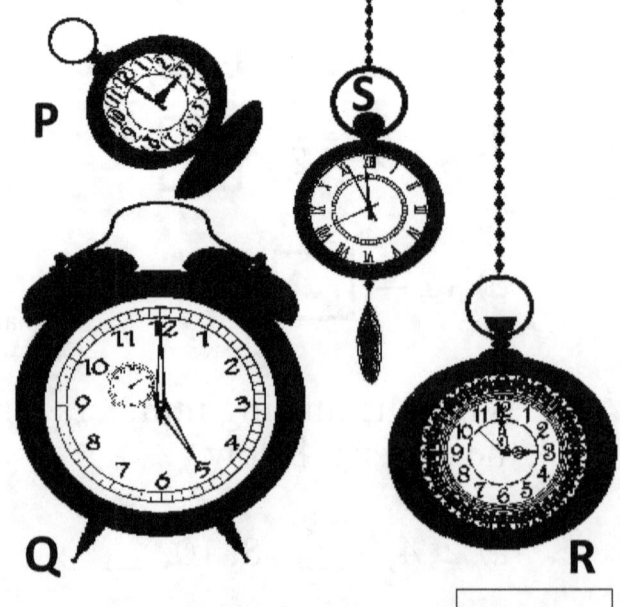

1 mark

10 Starting with the biggest, put these numbers in order of size

| 7 | 6 | 9 | 8 | 4 |

2 marks

1 Look at the shapes below.

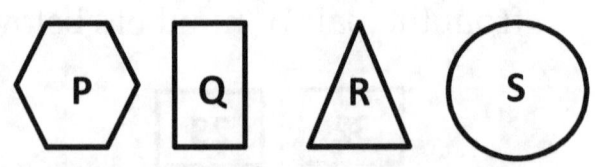

Choose from **triangle, rectangle, parallelogram, octagon and hexagon** to complete the following sentences.

a) Shape Q is a _____
b) Shape S is a _____
c) Shape P is a _____
d) Shape R is a _____

4 marks

12 Write the times for P,Q,R and S

End of Test

4 marks

Answers

PAGE 4
1)5   **2)**9   **3)**6   **4)**10   **5)**10   **6)**3   **7)**3   **8)**7
PAGE 5
1)1   2)2   3)2   4)4   5)3   6)3   7)4
8)7   9)8   10)7   11)0   12)6   13)8   14)6
PAGE 6
1) 10, red   2) 4, green   3) 20, purple   4) 20, purple
5) 4, green   6) 5, brown   7) 7, blue   8) 5, brown
9) 5, brown   10) 7, blue   11) 10, red   12) 4, green
PAGE 7
1) 1   2) 2   3) 3   4) 4   5) 5   6)6   7)7   8)8   9)9
10) 10   11) 2   12) 4   13)6   14) 8   15) 10   16) 12
17) 14   18) 16   19)18   20) 20   21) 2   22) 2   23) 2
24) 5   25) 4   26) 2   27) 1   28) 20
PAGE 8
1) One   2) five   3) ten   4) seventeen   5) twenty
6) 5 – five, 7 – seven, 10 – ten, 15 – fifteen, 20 - twenty
7) 8   8) 12   9) 20   10) 30   11) 35   12)53   13) 60
14) 66   15) 70   16) 79   17) 80   18) 87   19)90   20)100
PAGE 9
1) 2 2) 1 3) 6 4) 5 5) 11 6) 40 7) 50 8) 3.4 9) 8.0
10) 1.1   11) 12.5   12) 4.4   13) 34.4
PAGE 10
1a) 20   b) 20   c) 25   d) 12.   **2a)** 23   b) 28   c)23   d) 27
3a) 8   b) 5   c) 2   d) 6   e) 11   f) 11   g)30   h) 10   i) 10
j) 10   k) 50   L) 40   **4)** 20   **5a)** 60 pence   b) 40 pence   **6)** No
PAGE 11
1) 6,8,20 2) 3,7,9,11 3) 20,22,24,26,28 4a) 36 b)21
c) 15 5) Dad 6) 84 years 7) True 8) 31 9)False
PAGE 12
1a) 5,6 b) 10,12 c) 25,30 d) 11,13 e) 12,10 2a) 7,9
b) 20,50 c) 13,22 d) 13,21 e) 30,32 3) 18,22,26
4)Even 5a) ADD 3 b) ADD 2 c) Take away 3 d) ADD 5
6a)◯◯◯◯◯ ◯ b) 2 c)add 2 each time
7a) 18,28, b) ADD 5   8) take away 5   9b) 7,10,13
9c) 8,6,4   9d) 14,11,8
PAGE 14 1a)◯ b)△ c)⊕   2a) rectangle b) circle
c) square d) triangle e) pentagon f) hexagon
3) Hexagon⬡ Rectangle▭ Triangle△Pentagon⬠
Circle◯ Octagon⬡
PAGE 15 1a) Cylinder b) Cuboid c) Cone d) Pyramid
2a) ▭   b) ◯   3) No. Circle is 2D and sphere is 3D.
PAGE 16 1a) 6 b) 10 c) 13 d) 13 e) 9 f) 25
2a) 7 b) 0 c) 15 d) 20 e) 20 f) 20 g) 50
PAGE 17 1a) 50 pence b) £1 c) £4 2a) 30 pence
 b) £1.40 c) Yes, 4 rulers will cost £2.80. 3a) 8 b) 4
4a) 56 pence b) i) 1p, ii) 20p, 1p iii) 5p, 1p c) 4 pence
PAGE 18 1a) 47, b) 55 c) 76 d) 98 e) 97 f) 88
2a) 20 b) 80 c) 800 3) 48 4) 95 5) 90 6) 220 7) 86
8) 78 9) 44 10) 94 11) 99 12) 72
PAGE 19) 1) 61 2) 22 3) 52 4) 34 5) 82 6) 41 7) 10
8) 10 9)54 10) 20 11) 7 12)8
PAGE 20 1) 3,4,6,8,10 2) 3,8,9,10,19 3) 4,6,10,16,20
4) 14,24,40,41,56 5) 20,50,56,76,79 6) < 7) < 8) < 9)>
10) > 11) < 12) < 13) > **PAGE 21** 1) 12 2) 12:30 3) 1
4) 1:30 5) 2 6) 2:30 7) 3 8) 3:30 9) 4   10) 4:30 11)5
12) 5:30 13) 6 14) 6:30 15) 7 16) 7:30 17) 8 18) 8:30

19) 9   20) 9:30   21) 10   22) 10:30   23) 11   24) 11:30
PAGE 22

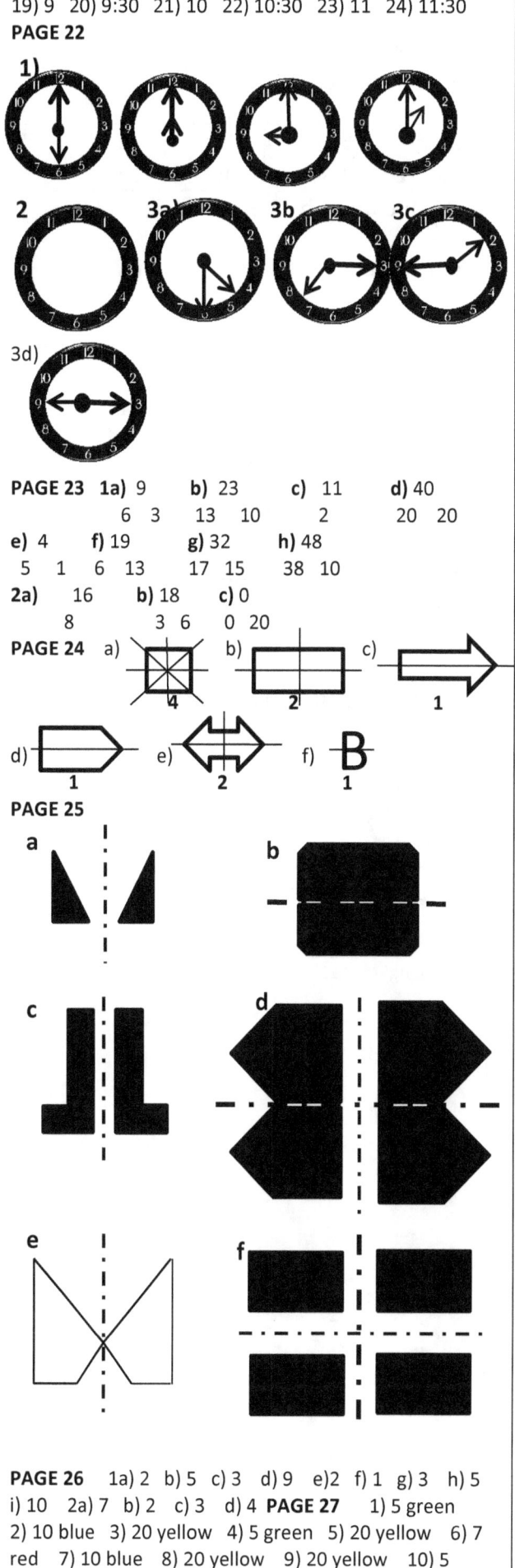

PAGE 23 **1a)** 9   **b)** 23   **c)** 11   **d)** 40
6   3   13   10   2   20   20
e) 4   **f)** 19   **g)** 32   **h)** 48
5   1   6   13   17   15   38   10
2a)   16   **b)** 18   **c)** 0
8   3   6   0   20
PAGE 24   a) b) c)
4   2   1
d) e) f) B
1   2   1
PAGE 25

PAGE 26   1a) 2 b) 5 c) 3 d) 9 e)2 f) 1 g) 3 h) 5
i) 10   2a) 7 b) 2 c) 3 d) 4 **PAGE 27**   1) 5 green
2) 10 blue 3) 20 yellow 4) 5 green 5) 20 yellow 6) 7
red 7) 10 blue 8) 20 yellow 9) 20 yellow 10) 5
green

PAGE 28

1) All answers ±2mm

Estimate	5cm	1.8cm	3.2cm	2.8cm
Exact L	4.9cm	2cm	3cm	3cm

2) a 10cm **b** 11.5cm **c** 6cm

3) a 100mm **b** 115mm **c** 60mm

PAGE 29

1) 50 2) 44 or 45 3a) 34°C b) 50 kg c) 1.5 kg d) 75 kg e) 40°C f) 129 kg

PAGE 30

1a

Colour	Tally	Frequency
Black	ЖТ	5
Grey	ЖТ //	7
Blue	///	3

b Grey **c** 15 pupils

2 a Mathematics **b** Science **c** 6

d 28 **e** Isabel is wrong. 6 + 10 = 16 and not 14

PAGE 31 - ASSESSMENT (TEST)

1a) 23 b) 2.7 **2)** 8,20,56 **3a)** 4 b) 1 c) 2 d)2 **4a)** 6, 12 b) 11, 5 **5a)** 10 kg b) 13 kg

6
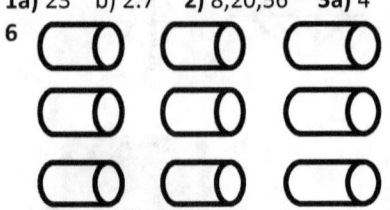

7) 33 **8a)** A = 3 b) B = 15 c) 80 **9)** A **10)** 9,8,7,6,4 **11a)** Rectangle b) Circle c) Hexagon
d) Triangle **12)** P : 3 o'clock Q : 5 o'clock R : 3 o'clock S) 11:55 or 5 to 12